A LOOK AT PHYSICAL SCIENCE

# HEAT



BY KATHLEEN CONNORS

Gareth Stevens PUBLISHING

CRASHCOURSE

**Please visit our website, www.garethstevens.com. For a free color catalog of all our high-quality books, call toll free 1-800-542-2595 or fax 1-877-542-2596.**

**Library of Congress Cataloging-in-Publication Data**

Names: Connors, Kathleen, author.
Title: Heat / Kathleen Connors.
Description: New York : Gareth Stevens Publishing, [2019] | Series: A look at physical science | Includes index.
Identifiers: LCCN 2017046113| ISBN 9781538221471 (library bound) | ISBN 9781538221495 (pbk.) | ISBN 9781538221501 (6 pack)
Subjects: LCSH: Heat--Juvenile literature. | Thermodynamics--Juvenile literature.
Classification: LCC QC256 .C657 2019 | DDC 536--dc23 LC record available at https://lccn.loc.gov/2017046113

Published in 2019 by
**Gareth Stevens Publishing**
111 East 14th Street, Suite 349
New York, NY 10003



Designer: Samantha DeMartin
Editor: Kristen Nelson

Photo credits: Series art Creative Mood/Shutterstock.com; cover, p. 1 weintel/Shutterstock.com; p. 5 (sauna) Przemyslaw Wasilewski/Shutterstock.com; p. 5 (sun) Aphelleon/Shutterstock.com; p. 5 (hot tub) Bill Kennedy/Shutterstock.com; p. 5 (campfire) prnkk/Shutterstock.com; p. 7 Vereshchagin Dmitry/Shutterstock.com; p. 9 gpointstudio/Shutterstock.com; p. 11 Karana/Shutterstock.com; p. 13 Ilya Marchenko/Shutterstock.com; p. 15 Narith Thongphasuk/Shutterstock.com; p. 17 Designua/Shutterstock.com; p. 19 Anna Jurkovska/Shutterstock.com; p. 21 Oliver Britton/Shutterstock.com; p. 23 corbac40/Shutterstock.com; p. 25 Samo Trebizan/Shutterstock.com; p. 27 LStockStudio/Shutterstock.com; p. 29 Romolo Tavani/Shutterstock.com; p. 30 (fire) Rvector/Shutterstock.com; p. 30 (oven mitts) Shahin Aliyev/Shutterstock.com; p. 30 (ice cubes) Great_Kit/Shutterstock.com; p. 30 (thermometer) EGOR_21/Shutterstock.com.



Printed in the United States of America

CPSIA compliance information: Batch #CS18GS: For further information contact Gareth Stevens, New York, New York at 1-800-542-2595.

# CONTENTS



Words in the glossary appear in **bold** type the first time they are used in the text.

# HEAT IS ENERGY

Heat takes many forms on Earth, from sunlight to hot water in the shower to the heat in an oven used to cook dinner. In all forms, heat is always energy! Heat is only one kind of energy at work on Earth.

### MAKE THE GRADE

Thermodynamics is the study of energy, such as heat, moving around **systems**.

# THE NEED FOR BALANCE

When energy moves from one system to another because of a temperature difference, that's heat! The highest temperature matter or area in a system is called the heat source. A lower temperature matter or area is called the heat sink.

## MAKE THE GRADE

Energy moves in order for a system to have **equilibrium**.

Heat moves from the heat source to the heat sink. This energy movement can't be seen because it happens between atoms and **molecules**. That means when you get warmer, you won't be able to see it happening!

### MAKE THE GRADE

All energy moves on the atomic level, not just heat.

# CREATING HEAT

The chemical reaction of burning makes heat. Another way heat is made is through **friction**. Heat is also made and lost when work is done **inefficiently**. This happens all the time, such as when a car engine runs or when **muscles** are used!

## MAKE THE GRADE

A chemical reaction is what happens when matter mixes with something else and it causes changes in the matter.

# MEASURE IT!

Temperature is a measurement that tells how hot or cold something is. It measures the speed at which the **particles** in matter are moving. The faster the particles are moving, the higher the temperature.

### MAKE THE GRADE

The slower the speed of matter's particles, the lower the temperature.

Thermometers measure temperature. They may have one or more **scales** written on them. Fahrenheit is the scale used in the United States. In Fahrenheit, water freezes at 32 degrees (°). Countries around the world use the Celsius scale. In Celsius, water freezes at 0°.

## MAKE THE GRADE

The unit scientists use to talk about an amount of heat is the calorie. A calorie is equal to the amount of heat energy needed for the temperature of 1 gram of water to rise 1° Celsius.

A common temperature scale used in science is Kelvin. The Kelvin scale is based on how much energy is in a system. A system with no **kinetic energy** is the coldest temperature possible. This is called absolute zero, and it's 0° on the Kelvin scale.

## MAKE THE GRADE

At absolute zero, there is no movement of atoms or molecules at all!

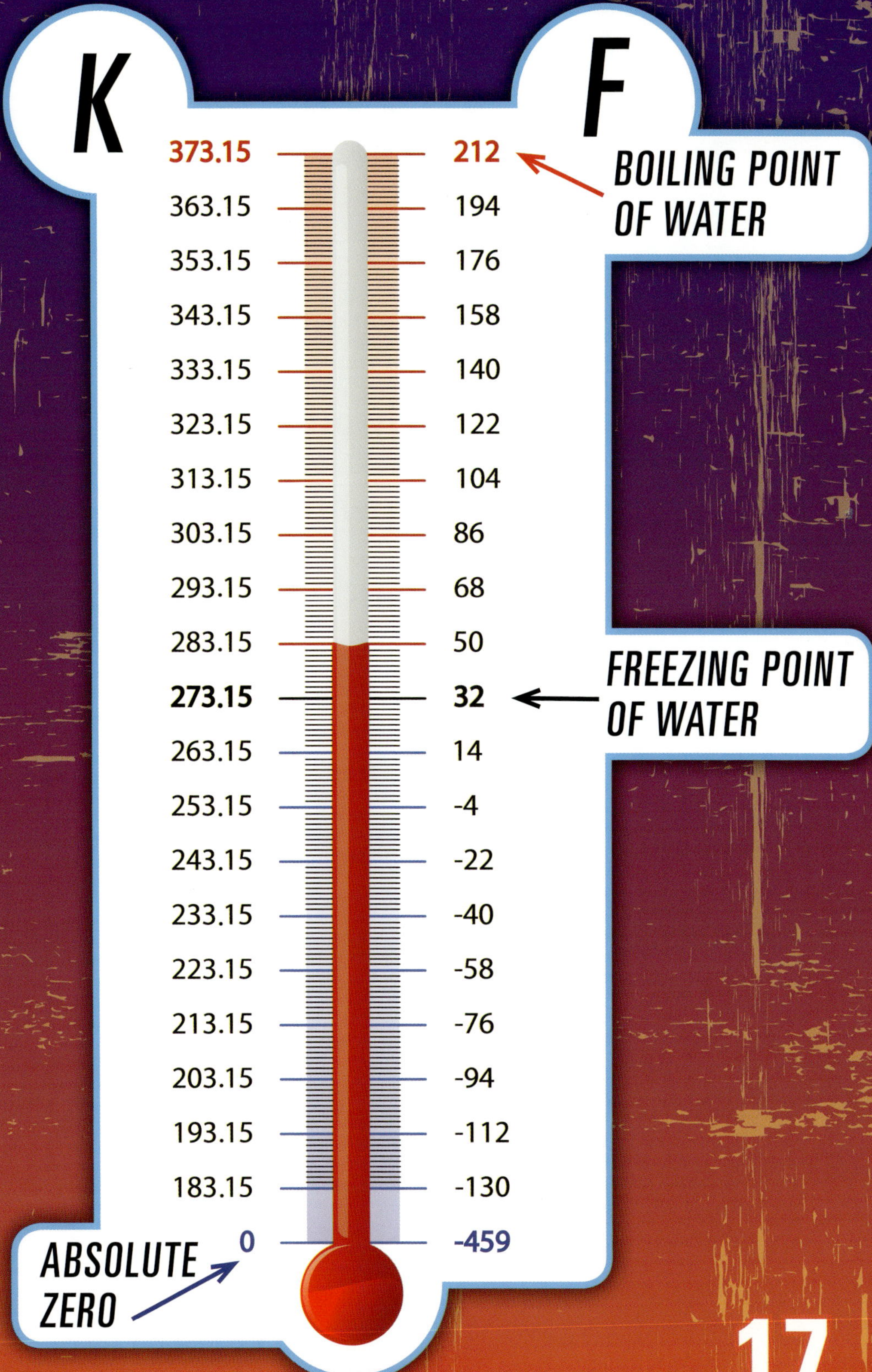

K
F
373.15
212
BOILING POINT OF WATER
363.15
194
353.15
176
343.15
158
333.15
140
323.15
122
313.15
104
303.15
86
293.15
68
283.15
50
273.15
32
FREEZING POINT OF WATER
263.15
14
253.15
-4
243.15
-22
233.15
-40
223.15
-58
213.15
-76
203.15
-94
193.15
-112
183.15
-130
0
-459
ABSOLUTE ZERO

# THE STATE OF MATTER

Heat can greatly affect matter. As temperature rises, gases, liquids, and solids expand, or get bigger. When matter loses enough heat, solids, liquids, and gases start to contract, or get smaller. Heat can also cause changes in the state of matter.

## MAKE THE GRADE

Gases expand and contract the most. Solids expand and contract the least.

If you add enough heat to a solid, it will become a liquid. If enough heat is added to a liquid, it will become a gas. When a gas cools, it becomes liquid. When a liquid cools, it becomes a solid.

## MAKE THE GRADE

The temperature at which it's hot enough for a solid to become a liquid is called the melting point.

# HEAT MOVEMENT AND TRANSFER

Convection is the movement of large amounts of matter into a system that has a different temperature. The system gets warmer or cooler because of this movement. Convection occurs only in liquids and gases. It happens in Earth's **atmosphere** and oceans all the time!

### MAKE THE GRADE

A convection oven uses a fan to move warm air around to help food cook more evenly.

Conduction is the transfer of heat energy when matter touches. The heat moves from the matter with the higher temperature to the matter with the lower temperature. The heat moves until all the matter is the same temperature.

## MAKE THE GRADE

In order to keep heat from moving from one system to another, people use insulators. Like feathers or fur on animals, insulators are matter that stop or slow heat transfer.

INSULATORS

# HEAT RADIATION

In another type of heat transfer, there is no contact between matter. Radiation is heat leaving matter as electromagnetic waves. These waves are the energy and light that move around us all the time. We can only see some types of electromagnetic waves.

## MAKE THE GRADE

Everything radiates energy all the time. The more heat something has, the more energy it gives off!

Everything radiates heat—including you! The higher the temperature of the matter, the stronger the heat radiation will be. One of the best examples of radiation is the heat of the sun reaching Earth!

### MAKE THE GRADE

Radiation also causes matter to cool. This happens because heat energy always moves toward matter or systems of lower amounts of heat energy.

# HEAT BASICS

Heat is a form of energy.

Temperature is a measure of heat. It can be written in units called Celsius, Fahrenheit, or Kelvin.

Heat energy always moves from matter with higher temperatures to matter with lower temperatures.

Gaining or losing heat can cause matter to change states.

Heat is made by friction, chemical reactions, or when work is done.

Insulators can slow or stop heat transfer.

Heat moves or is transferred by conduction, convection, and radiation.

# GLOSSARY

**atmosphere:** the mixture of gases that surround a planet

**equilibrium:** a state of balance

**friction:** the force that slows motion between two objects touching each other

**inefficiently:** having to do with the least effective or purposeful way of doing something

**kinetic energy:** the energy having to do with movement

**particle:** a very small part of something

**molecule:** a very small piece of matter

**muscle:** one of the parts of the body that allow movement

**scale:** a range of numbers used to show the size or strength of something

**system:** a group of connected parts that work together

# FOR MORE INFORMATION

## BOOKS

Polinsky, Paige V. *Super Simple Experiments with Heat and Cold: Fun and Innovative Science Projects.* Minneapolis, MN: Super Sandcastle, 2017.

Steinberg, Lynnae D. *What Is Heat Energy?* New York, NY: Britannica Educational Publishing, 2018.

## WEBSITES

**Heat Facts**

*www.scienceforkidsclub.com/heat-facts.html*

Read more about heat on this fun website.

**Physics for Kids**

*www.sciencekids.co.nz/physics.html*

Find out about many more physical science topics here.

**Publisher's note to educators and parents:** Our editors have carefully reviewed these websites to ensure that they are suitable for students. Many websites change frequently, however, and we cannot guarantee that a site's future contents will continue to meet our high standards of quality and educational value. Be advised that students should be closely supervised whenever they access the internet.

# INDEX